Copybook II

Copy Book II

ISBN #: 1-930953-86-0
Memoria Press
www.memoriapress.com

Table of Contents

INTRODUCTION

What is copybook?

Copybook is a time-honored learning activity in which students copy Scripture, maxims, poetry and other literature selections. Copybook, like memorization, is a forgotten skill in modern education. However, like memorization, copybook is a beginning skill that is indispensable to the development of good language skills.

What important skills are practiced in copybook? In Copybook students practice penmanship, spelling, reading comprehension, punctuation, and vocabulary. Students develop the habits of accuracy, neatness, attention to detail, and patience; they practice correct grammar and good writing style; they develop an appreciation for what is good, noble, and beautiful in literature.

Copybook is not the only way for the young student to practice each of these skills - but it is the only exercise that develops all of them at the same time. Copybook is concrete and physical; it complements and completes the mental exercise of memorization and reading. Copybook is so simple and obvious that we overlook it.

Even though penmanship is hard work for the young student, he should come to delight in the physical act of creating a beautiful page of writing. Because the student is not burdened by trying to compose his own words, he is liberated to concentrate on copying the immortal words of others. The young student who is not able to think great thoughts of his own, can grow in wisdom by thinking great thoughts of others. And as a reward for completing his copybook exercise he can illustrate the words he has copied with his own drawing. After the discipline of copybook he can exercise his imagination and creativity; the original drawings of children are the perfect companion to copybook.

How did we select our Bible verses?

The King James Bible is the poetic and literary version of Scripture. It has had a tremendous influence on the development of the English language. Many spiritual and literary allusions and expressions come directly from the King James Bible. A thorough knowledge of the King James Bible is a necessary preparation for the study of English literature, as well as a foundation for a deep spiritual life.

Because memorization is a lost art, parents and teachers today rarely have a vision for the quantity and quality of Scripture and poetry that children can and should commit to memory. The selection of passages is important in that they should lead the young child into a love and appreciation for Scripture; they should be age-appropriate passages, ones that employ poetic language to paint concrete pictures of Bible stories. Verses in our K-2 Copybooks focus on the great, dramatic, stories in salvation history rather than on doctrinal or abstract concepts, which are best reserved for the more mature student.

The K-2 Copybook memory verses are reviewed in the 3-year Christian Studies program for grades 3-5. The goal is for students to memorize and retain all of the Scripture they have learned over the course of their entire school years so that each year they can recite all of the verses from the current year plus all from previous years. A complete Bible memorization program allows for the student to absorb a vast body of Scripture of great variety and content over the whole of the Bible.

TEACHING GUIDELINES

The first task of the teacher is to communicate the importance of the lesson to the student and to model enthusiasm. The child should see copybook as an opportunity to give his very best effort in penmanship and artwork throughout the year. Students should be inspired to produce quality work because the words they are asked to memorize and copy are the sacred words of Scripture. The quality of the work will be a direct reflection of the ability of the teacher to motivate the child to excellence.

Beginning kindergarten students have varying skills in reading and printing, but most know some letters and sounds. Even though students may not be able to read the verse for the first half of kindergarten, they can still memorize and copy the words. As their reading skills grow throughout the year, they will be able to review their verses with increasing understanding, and look back on their improving penmanship.

The Copybook Lesson outlined below need not be done in one sitting, but may be broken up into two or more sessions. There are 7 skills or steps for each lesson.

Step 1 ______________________________

Bible Story Time. Write the memory verse on the board. Read the verse from a Bible. A traditional soft leather Bible is best because it communicates better that the Bible is a special book, the Sacred Word of God. Keep the Bible in your room or home in a special place and handle it with reverence. You will want to read the context of the verse from the Bible and/or from a story book. We recommend the Golden Children's Bible as a story book because its words are closest to Scripture and the pictures are realistic and memorable.

Step 2 ______________________________

Language Lesson. Using the blackboard, study the verse with students. Identify unfamiliar words and explain their meanings in context, rather than by using the dictionary. Scripture and poetry often contain archaic or poetic words that are not common in contemporary writing, words such as thee, thy, giveth, firmament, etc. Children are intrigued by these 'interesting words' and early exposure to them enriches their language education. Even though Scripture and poetry may also have non-standard punctuation and grammar, the passages can be used to introduce simple grammar topics (nouns, verbs), punctuation, capitalization, plurals, contractions, etc. Kindergarten students, of course, will not be ready for these topics until they are reading, usually the second half of the kindergarten year.

Step 3 ______________________________

Memorization. Use the disappearing verse technique. Recite the whole verse together several times. Break the verse down into small sections. Have students recite one section together, and then erase that section. Pointing to the 'erased' section on the board, ask students individually and then in unison, to recite what you erased. Repeat until the whole verse has been erased from the board. (Knowing the words are going to be erased helps students concentrate and adds excitement.) By the end of the process, students will be able to repeat the whole verse from memory. For instance for the first passage the three sections would be:

And God said
Let there be light
And there was light.

Step 4 ___

Copying. Now that students are familiar with the verse, they copy it in their best handwriting. It is critical that students develop correct techniques in holding the pencil and forming letters. The teacher should monitor students during copybook time and quickly correct any errors in technique. The great value of copybook is that unlike all of the other writing students do, copybook is a special time for students to give their very best effort. **Some first grade students may not be able to complete copy verses at the beginning of the year. They should memorize and illustrate verses according to the schedule but work on alphabet pages for step 4 until they are ready to do copy work.*

Step 5 ___

Proofreading and Correction. Students should proof and correct all work. This is a critical skill that must be taught, encouraged and practiced from the beginning of the child's education. The proofreading should be kept as a separate exercise and sufficient time devoted to it. The lesson is not complete until this exercise of proofreading and correction produces a 'finished' product.

Step 6 ___

Illustration. Students can now draw a picture to illustrate the Bible verse or poem in the box below each passage. Help students to include as many details as possible. Art brings the copybook exercise to a close with delight and imagination. These drawings will be cherished in years to come.

Step 7 ___

Review. Part of every day should be devoted to review of verses learned. Students love to practice what they already know. Give students the first word or two of each passage and let them complete the passage as a group recitation or individually. As the verses accumulate, students will be amazed at what they know and will develop confidence in their abilities. At the end of the year students should be able to recite all verses with fluency.

Recommended Schedule

Please see complete instructions for each step on previous pages

Day 1: Step One - Bible Story Time (15 min)

Day 2: Step Two - Language Lesson (15 min.)

Day 3: Steps Three & Four - Memory Work and Copying (20 min.)

Day 4: Steps Five & Seven - Proofreading and Cumulative Review of all Memory Work (15-30 min)

Day 5: Step Six - Illustration (20 min.)

Week 1 .... Genesis 1:16
Week 2 .... Genesis 1:31
Week 3 .... Genesis 3:7
Week 4 .... Genesis 3:13
Week 5 .... Genesis 3:20 & Genesis 4:9
Week 6 .... Genesis 9:13
Week 7 .... Genesis 37:19
Week 8 .... Exodus 3:8
Week 9 .... Exodus 23:20
Week 10 .... Proverbs 17:22 & Proverbs 25:11
Week 11 .... Luke 2:12
Week 12 .... Acts 20:35 & I Corinthians 3:17
Week 13 .... Revelation 3:20
Week 14 .... Twelve Apostles
Week 15 .... Psalm 100
Week 16 .... Psalm 100
Week 17 .... Doxology
Week 18 .... Humilty
Week 19 .... Tired Tim
Week 20 .... Tired Tim
Week 21 .... Mr. Meant To
Week 22 .... Mr. Meant To
Week 23 .... January Brings the Snow
Week 24 .... January Brings the Snow
Week 25 .... January Brings the Snow
Week 26 .... January Brings the Snow
Week 27 .... Frogs at School
Week 28 .... Frogs at School
Week 29 .... Frogs at School
Week 30 .... Frogs at School

Line and Letter Practice

We have found that little instruction is necessary to teach basic letter strokes. Instead, it is "practice, practice, practice" on words and passages that creates neat and proficient writers. The following page, however, can be used to introduce letter formation and reinforce areas that prove problematic. You may use these exercises to learn and practice strokes and letters before beginning copywork. If your student does not need instruction on creating individual letters, you may skip this section and move straight to the first Bible verse.

Letter Practice Overview

Start with basic vertical, horizontal, slanted, and circular strokes. Do not underestimate a young student's ability to create simple strokes and stay within the lines. Always encourage straight lines, round circles, and appropriately sized letters. Once the basic strokes are mastered, your student should be able to make every letter in the alphabet. Each letter includes numbered stroke instruction. Many letters are created in a single continuous stroke where the student retraces his previous path.

Examples: n p

Other letters require two or more strokes and each stroke will be numbered in sequence:

Examples: f I

It is not necessary to practice each letter over and over until it is perfect, just until your student has the basic ability to form letters correctly. The purpose of Copybook is to allow your student to practice handwriting on great passages, not lists of letters. You will notice significant improvements in letter formation as the year progresses.

Remember, there are several different methods of creating letters. If your child has learned another system that works for him, skip the letter instruction section and move right into the copywork.

Student Guidelines

1. Students should hold pencils properly.

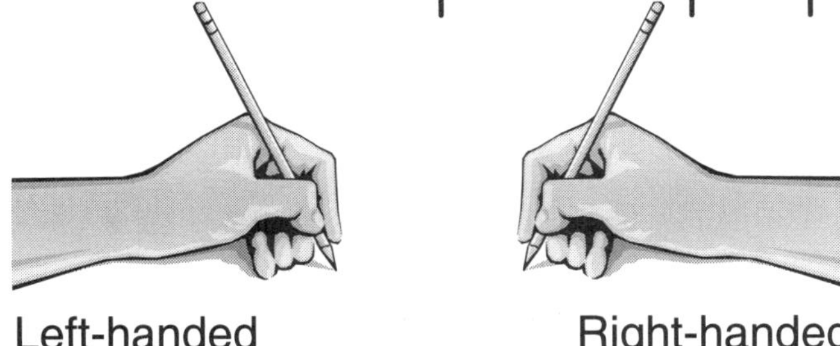

Left-handed Right-handed

2. Start each line close to the left margin and don't cross the right margin.
3. Uppercase letters must touch the top and bottom lines.
4. Most lower case letters stop at the middle, dashed line.
5. Lower case b, d, f, h, k, and l are all tall letters that touch the top line.
6. Lower case g, j, p, q, and y have hooks or tails that cross the bottom line.
7. All words should be one pinky space apart.
8. Letters should all stand tall and straight.
9. Erase mistakes completely and be neat.

Trace & Copy Horizontal & Vertical Lines

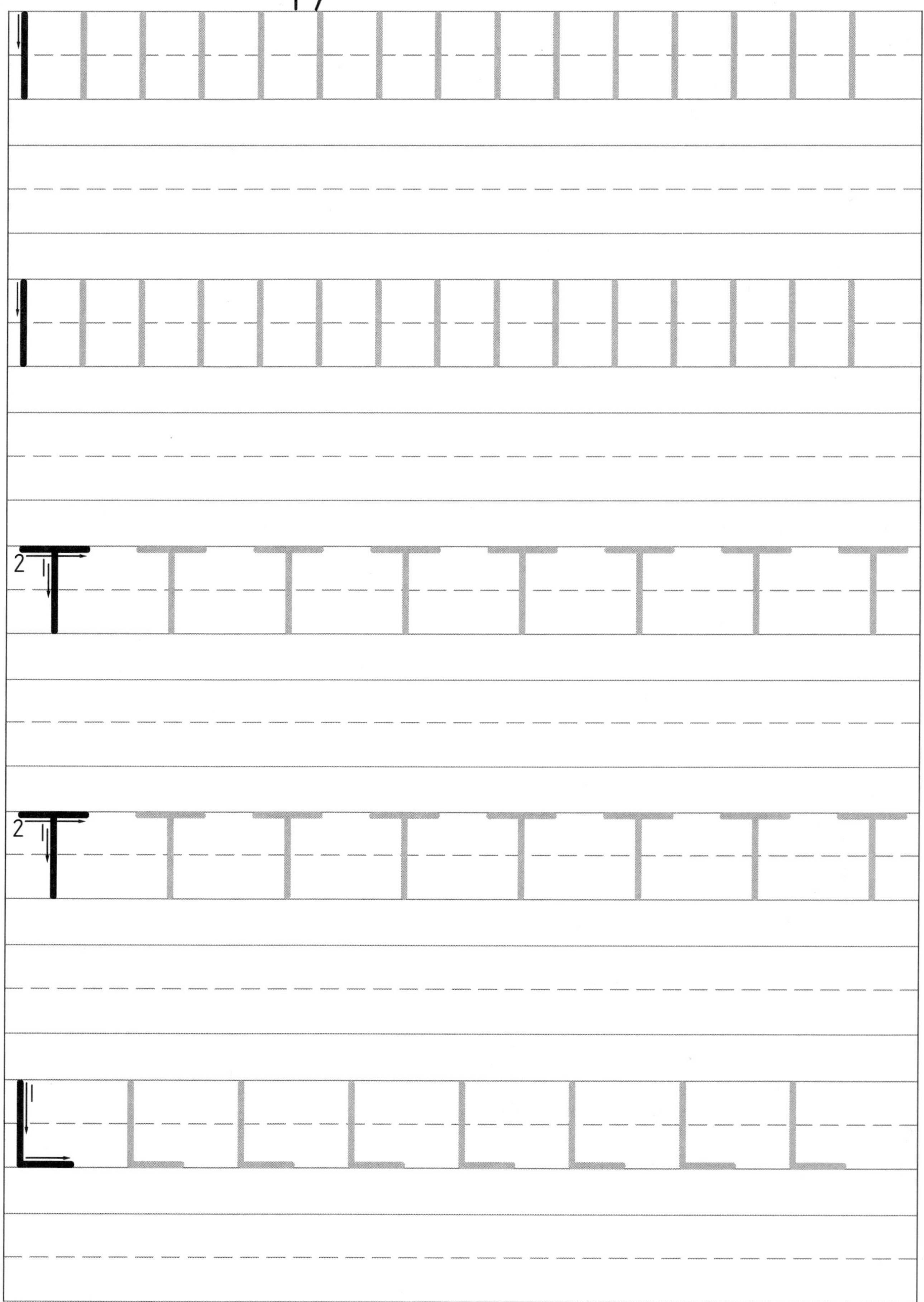

Trace & Copy Horizontal & Vertical Lines

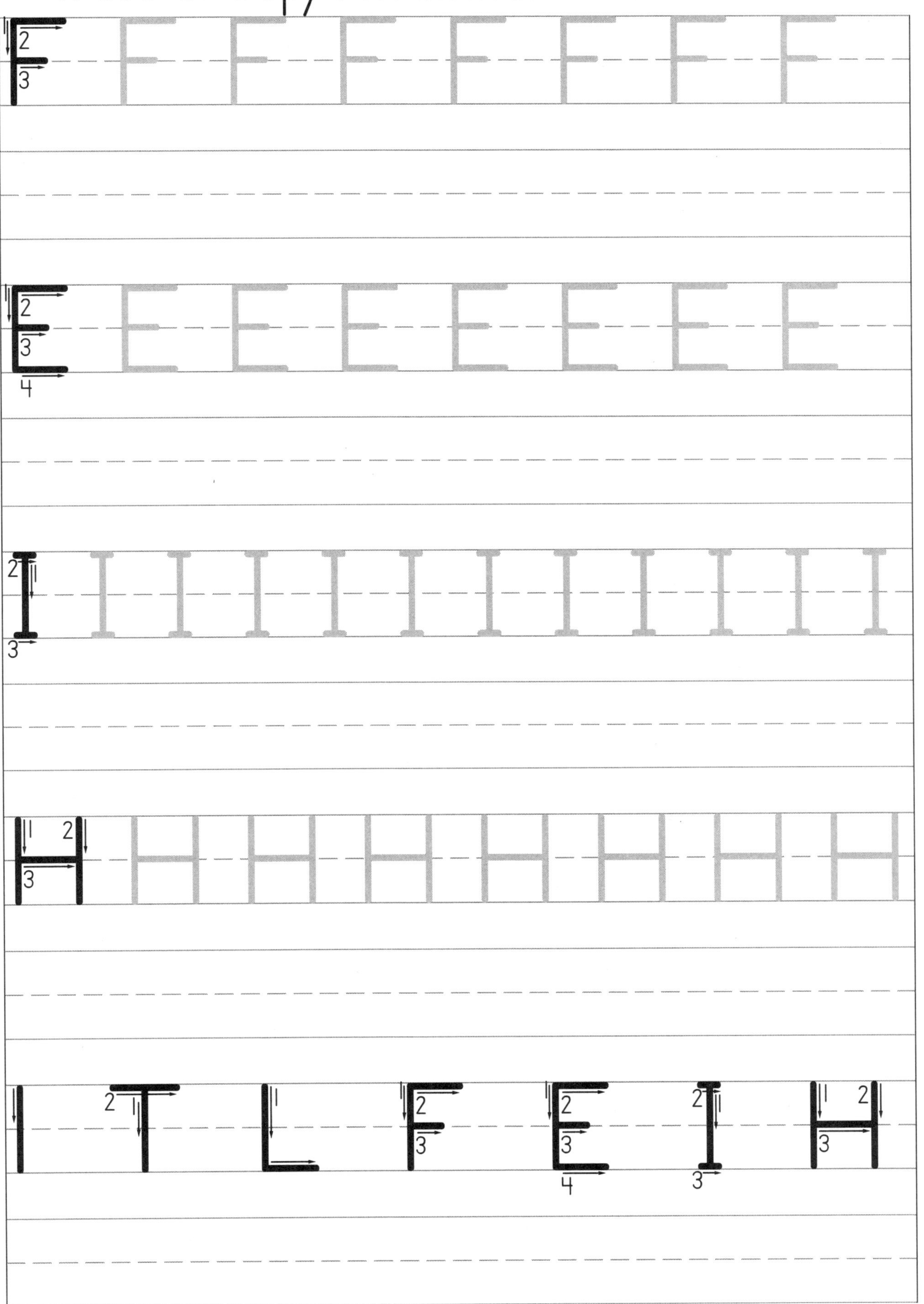

Trace & Copy Slanted Strokes

V V V V V V V V

X X X X X X X

Z Z Z Z Z Z Z Z

Y Y Y Y Y Y Y

N N N N N N N

Trace & Copy Slanted Strokes

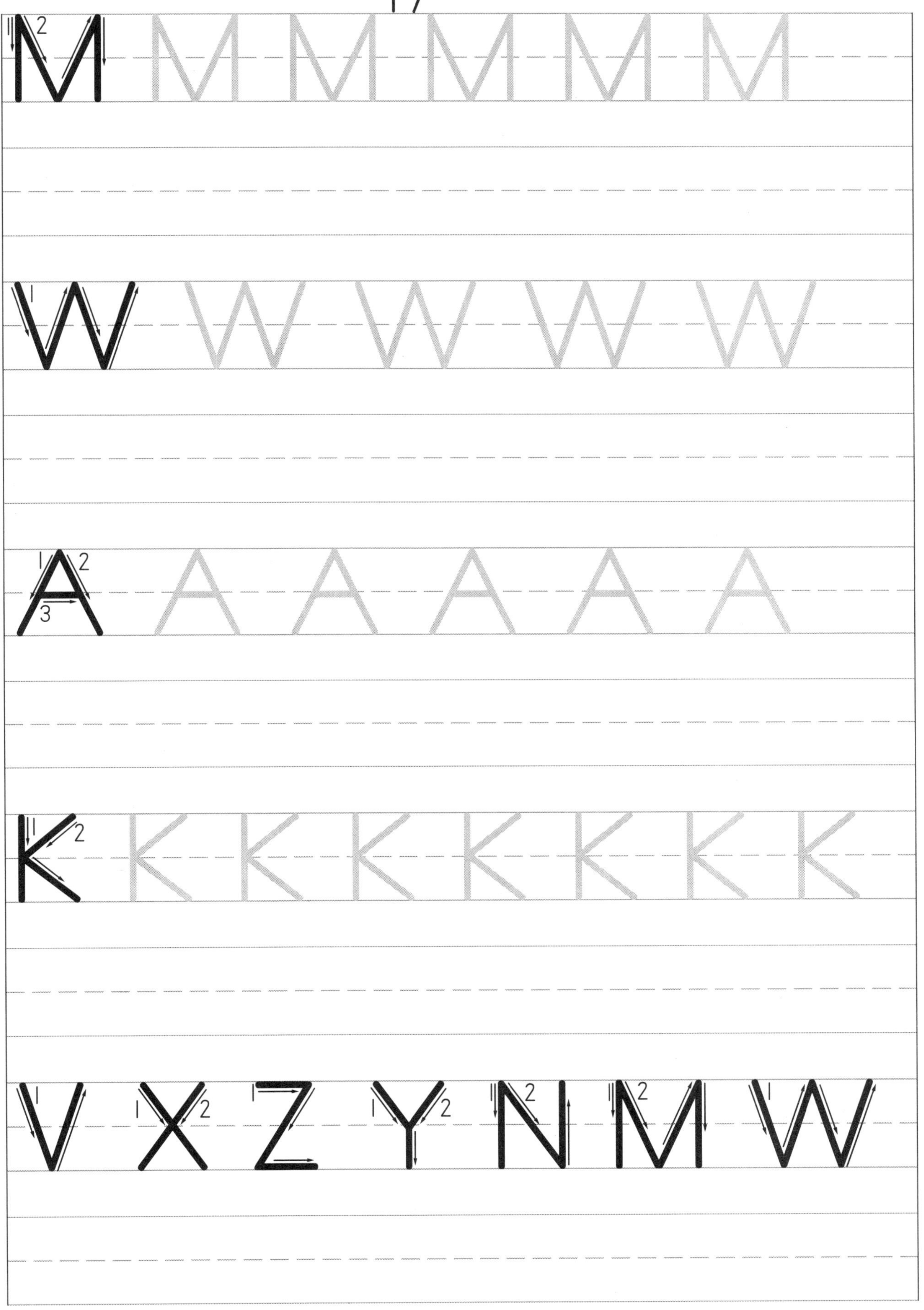

Trace & Copy Round Strokes

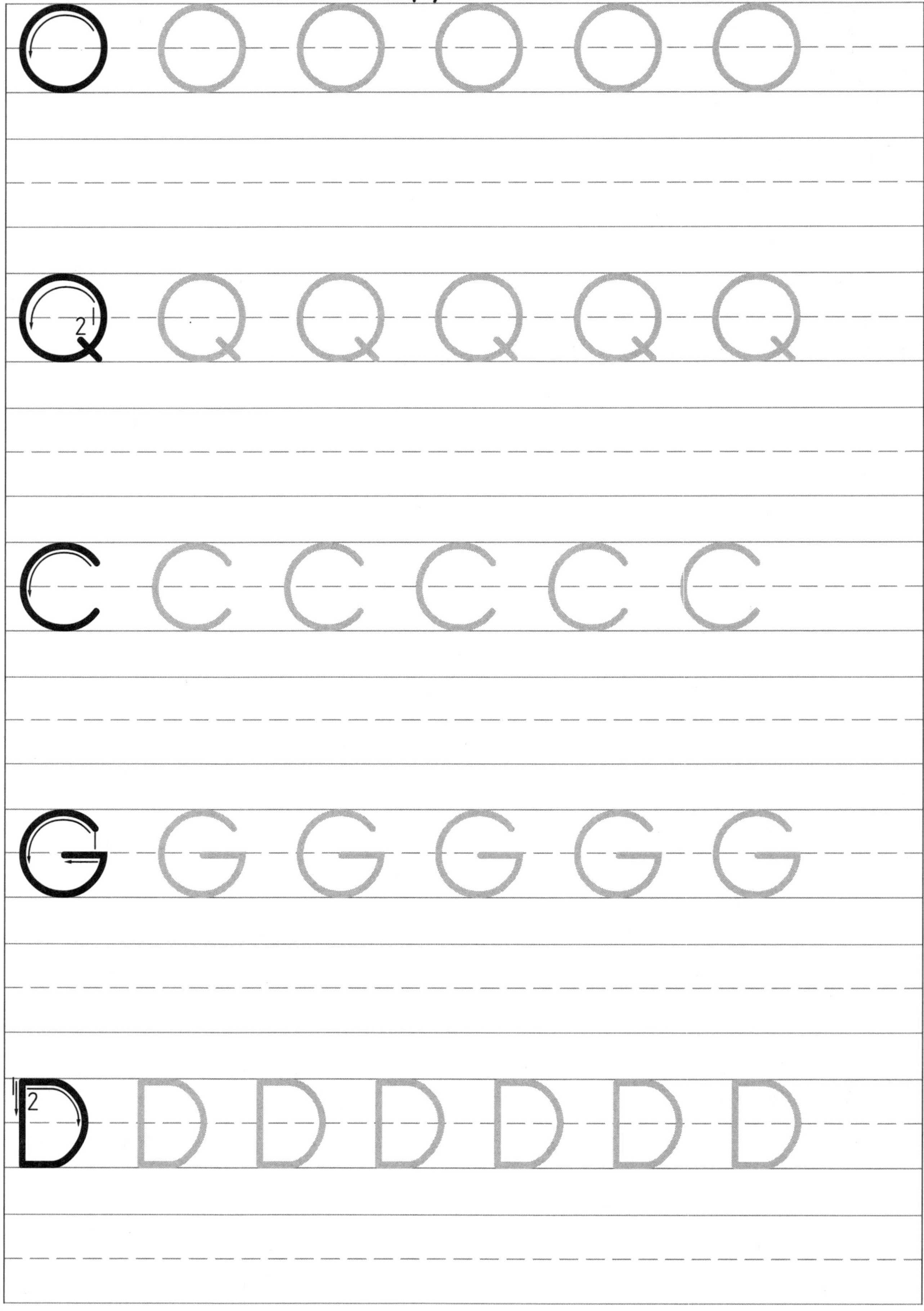

Trace & Copy Round Strokes

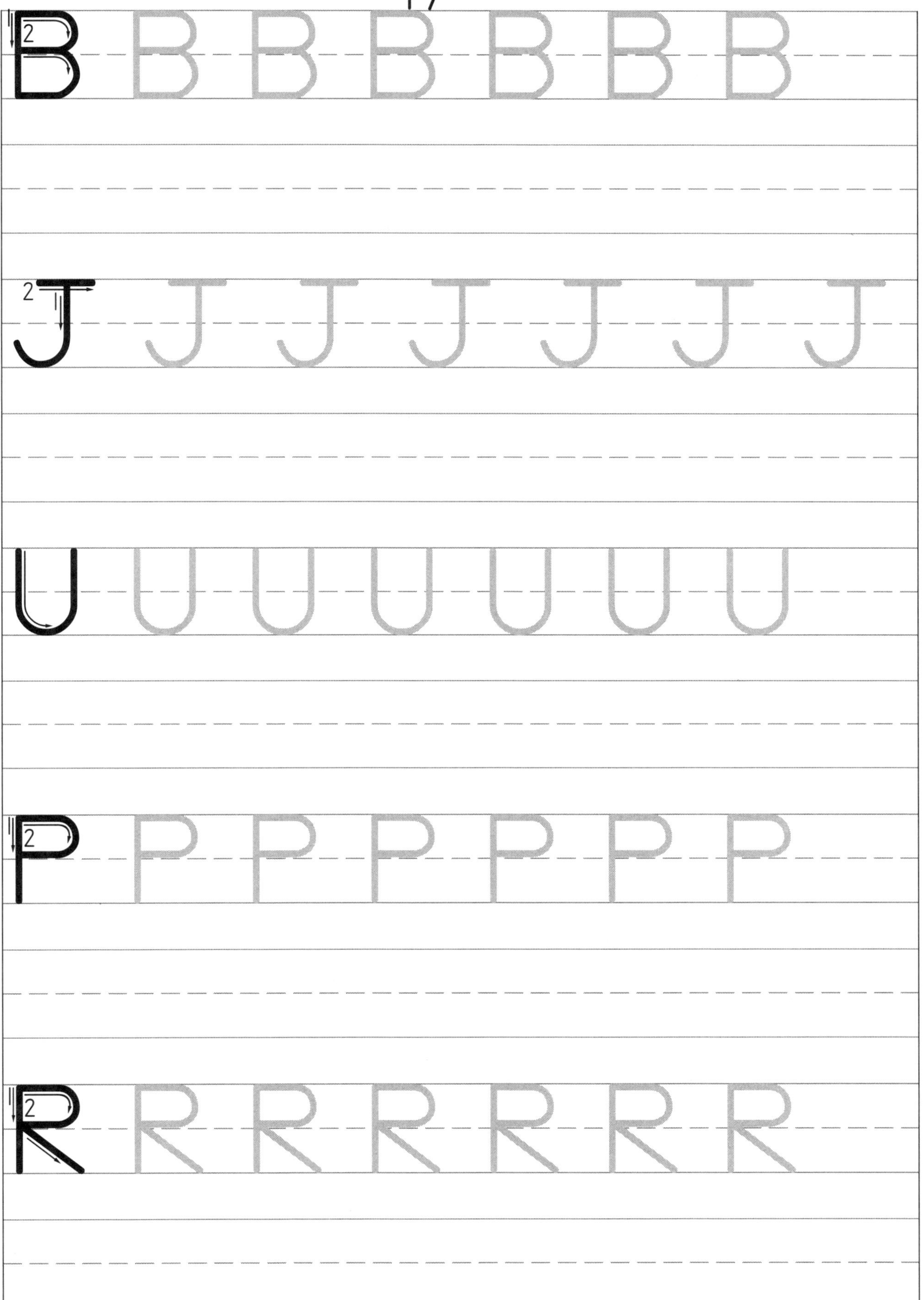

Trace

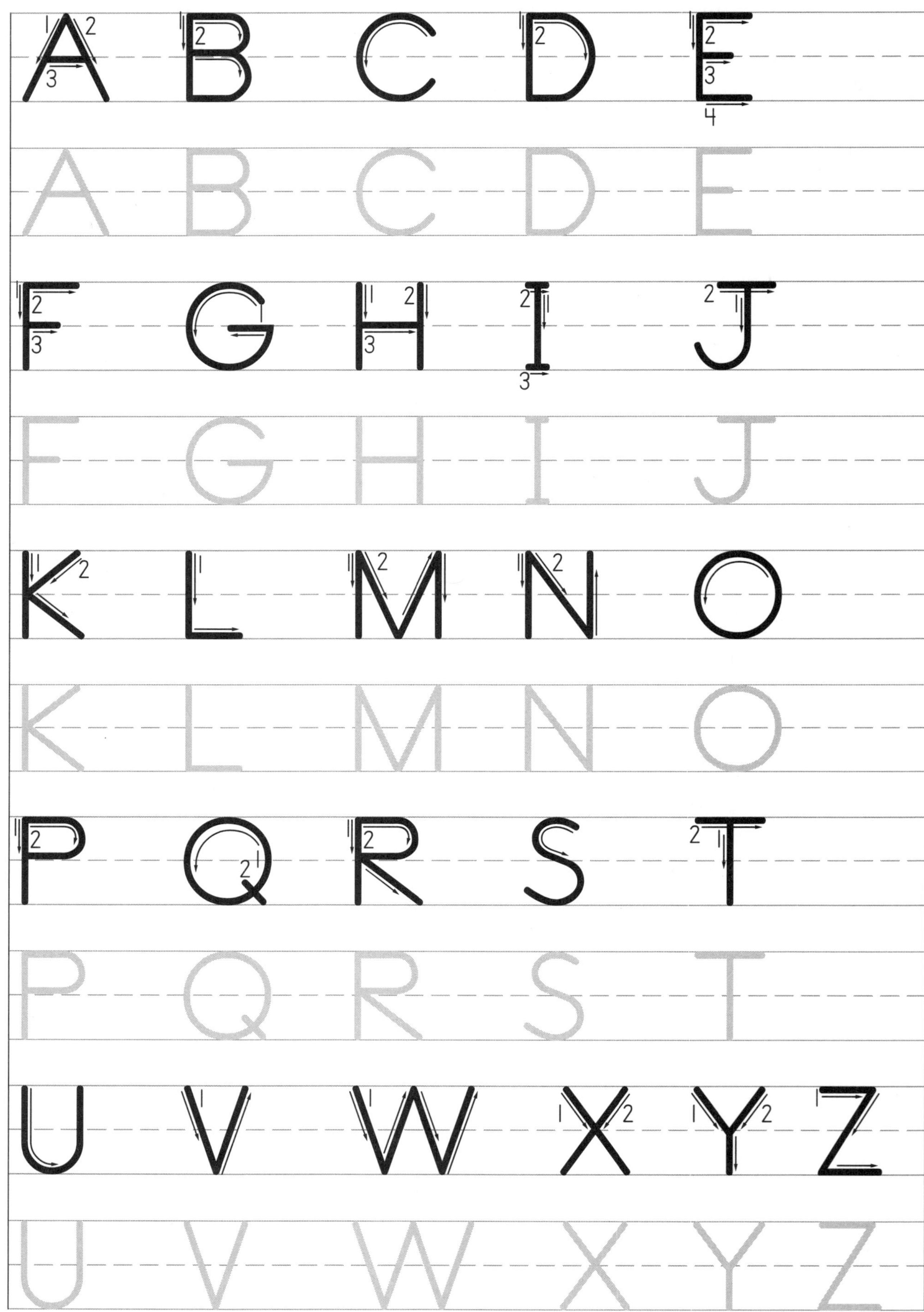

Trace and Copy

Copy

A B C D E

F G H I J

K L M N O

P Q R S T

U V W X Y Z

Copy From the Previous Page

Trace

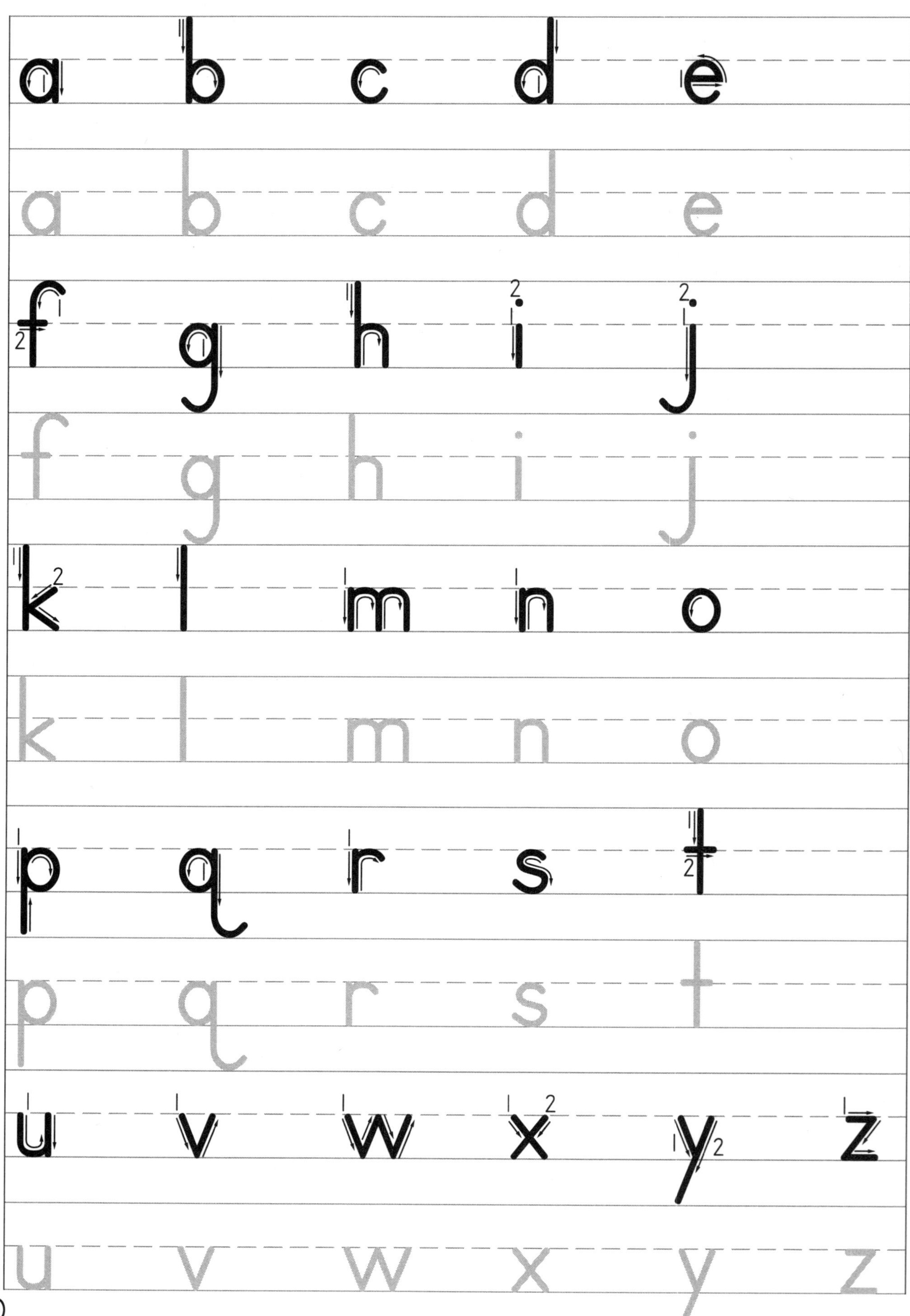

Trace and Copy

a b c d e

f g h i j

k l m n o

p q r s t

u v w x y z

Copy

a b c d e

f g h i j

k l m n o

p q r s t

u v w x y z

Copy From the Previous Page

Copy

Aa Bb Cc Dd Ee

Ff Gg Hh Ii Jj Kk

Ll Mm Nn Oo Pp

Qq Rr Ss Tt Uu

Vv Ww Xx Yy Zz

Copy From the Previous Page

And God made two great lights;
the greater light to rule the day,
and the lesser light to rule the night:
He made the stars also.

Genesis 1:16

And God made

two great lights;

the greater light

to rule the day,

And God made two great lights;
the greater light to rule the day,
and the lesser light to rule the night:
He made the stars also.

Genesis 1:16

and the lesser

light to rule the

night: He made

the stars also.

Genesis 1:16

And God saw every thing
that He had made,
and, behold, it was very good.

Genesis 1:31

And God saw

every thing that

He had made, and,

behold, it was very

good. Genesis 1:31

And they heard the voice
of the Lord God
walking in the garden
in the cool of the day.

Genesis 3:7–8

And they heard the

voice of the Lord

God walking in the

garden in the cool of

the day. Genesis 3:7-8

And the woman said,
The serpent beguiled me,
and I did eat.

Genesis 3:13

And the woman
said, The serpent
beguiled me, and I
did eat.

Genesis 3:13

And Adam called his wife's name Eve, because she was the mother of all living.

Genesis 3:20

And Adam called

his wife's name Eve,

because she was the

mother of all living.

Genesis 3:20

Am I my brother's keeper?

Genesis 4:9

Am I my brother's keeper?

Genesis 4:9

I do set my bow in the cloud,
and it shall be for a token
of a covenant
between me and the earth.

Genesis 9:13

I do set my bow in

the cloud, and it shall

be for a token of a

covenant between

me and the earth.

And they said to one another,
behold this dreamer cometh!

Genesis 37:19

And they said to

one another, behold

this dreamer

cometh!

Genesis 37:19

A land flowing with
milk and honey.

Exodus 3:8

A land flowing

with milk and

honey.

Exodus 3:8

Behold, I send an Angel before thee,
to keep thee in the way.

Exodus 23:20

Behold, I send an

Angel before thee,

to keep thee in the

way.

Exodus 23:20

A merry heart doeth good like medicine.

Proverbs 17:22

A merry heart

doeth good like

medicine.

Proverbs 17:22

A word fitly spoken is like apples of gold in pictures of silver.

Proverbs 25:11

A word fitly

spoken is like

apples of gold in

pictures of silver.

Proverbs 25:11

For unto you is born
this day in the city of David
a Savior, which is Christ the Lord.

Luke 2:12

For unto you is
born this day in
the city of David
a Savior, which is
Christ the Lord.

It is more blessed to give than to receive.

Acts 20:35

It is more blessed

to give than to

receive.

Acts 20:35

The temple of God is holy,
which temple ye are.

I Corinthians 3:17

The temple of

God is holy, which

temple ye are.

I Corinthians 3:17

Behold, I stand at the door, and knock.

Revelations 3:20

Behold, I stand

at the door, and

knock.

Revelations 3:20

The Twelve Apostles:

Peter
Andrew
James
John
Philip
Thomas

Matthew
Bartholomew
James the less
Simon
Thaddeus
Judas

The Twelve Apostles

Peter, Andrew,

James, John,

Philip, Thomas,

The Twelve Apostles:

Peter	Matthew
Andrew	Bartholomew
James	James the less
John	Simon
Philip	Thaddeus
Thomas	Judas

Matthew,

Bartholomew,

James the less,

Simon, Thaddeus,

Judas

Make a joyful noise unto the Lord, all ye lands.
Serve the Lord with gladness;
come before His presence with singing.
Know ye that the Lord He is God;
it is He that hath made us,
and not we ourselves;
we are His people,
and the sheep of His pasture.

Psalm 100

Make a joyful

noise unto the

Lord, all ye lands.

Serve the Lord

with gladness;

Make a joyful noise unto the Lord, all ye lands.
Serve the Lord with gladness;
come before His presence with singing.
Know ye that the Lord He is God;
it is He that hath made us,
and not we ourselves;
we are His people;
and the sheep of His pasture.

Psalm 100

come before His

presence with

singing.

Know ye that the

Lord He is God;

Make a joyful noise unto the Lord, all ye lands.
Serve the Lord with gladness;
come before His presence with singing.
Know ye that the Lord He is God;
it is He that hath made us,
and not we ourselves;
we are His people,
and the sheep of His pasture.

Psalm 100

it is He that

hath made us,

and not we

ourselves;

Make a joyful noise unto the Lord, all ye lands.
Serve the Lord with gladness;
come before His presence with singing.
Know ye that the Lord He is God;
it is He that hath made us,
and not we ourselves;
we are His people,
and the sheep of His pasture.

Psalm 100: 1–3

we are His people,

and the sheep of

His pasture.

Psalm 100: 1-3

Doxology

Glory be to the Father
and to the Son and to the Holy Spirit.
As it was in the beginning,
is now and ever shall be,
world without end.
Amen.

Doxology

Glory be to the

Father and to the

Son and to the

Holy Spirit.

Doxology

Glory be to the Father
and to the Son and to the Holy Spirit.
As it was in the beginning,
is now and ever shall be,
world without end.
Amen.

As it was in the

beginning, is now

and ever shall be,

world without end.

Amen.

Humility

Humble we must be,
If to Heaven we go.
High is the roof there,
But the gate is low.

Humble we must be,

If to Heaven we go.

High is the roof

there, But the

gate is low.

Tired Tim

Walter De La Mare

Poor Tired Tim! It's sad for him.
He lags the long bright morning through,
Ever so tired of nothing to do;
He moons and he mopes the livelong day,
Nothing to think about, nothing to say;
Up to bed with his candle to creep,
Too tired to yawn, too tired to sleep:
Poor tired Tim! It's sad for him.

Poor Tired Tim!

It's sad for him.

He lags the long

bright morning

through,

Tired Tim

Walter De La Mare

Poor Tired Tim! It's sad for him.
He lags the long bright morning through,
Ever so tired of nothing to do;
He moons and he mopes the livelong day,
Nothing to think about, nothing to say;
Up to bed with his candle to creep,
Too tired to yawn, too tired to sleep:
Poor tired Tim! It's sad for him.

Ever so tired of

nothing to do;

He moons and he

mopes the livelong

day,

Tired Tim

Walter De La Mare

Poor Tired Tim! It's sad for him.
He lags the long bright morning through,
Ever so tired of nothing to do;
He moons and he mopes the livelong day,
Nothing to think about, nothing to say;
Up to bed with his candle to creep,
Too tired to yawn, too tired to sleep:
Poor tired Tim! It's sad for him.

Nothing to think

about, nothing to

say; Up to bed

with his candle to

creep,

Tired Tim

Walter De La Mare

Poor Tired Tim! It's sad for him.
He lags the long bright morning through,
Ever so tired of nothing to do;
He moons and he mopes the livelong day,
Nothing to think about, nothing to say;
Up to bed with his candle to creep,
Too tired to yawn, too tired to sleep:
Poor tired Tim! It's sad for him.

Too tired to yawn,

too tired to sleep:

Poor tired Tim!

It's sad for him.

Mr. Meant-To

Anonymous

Mr. Meant-To has a comrade,
And his name is Didn't-Do.
Have you ever chanced to meet them?
Did they ever call on you?

These two fellows live together,
In the house of Never-Win.
And I'm told that it is haunted,
By the ghost of Might-Have-Been.

Mr. Meant-To

has a comrade,

And his name is

Didn't-Do.

Mr. Meant-To

Anonymous

Mr. Meant-To has a comrade,
And his name is Didn't-Do.
Have you ever chanced to meet them?
Did they ever call on you?

These two fellows live together,
In the house of Never-Win.
And I'm told that it is haunted,
By the ghost of Might-Have-Been.

Have you ever chanced to meet them?

Did they ever call on you?

Mr. Meant-To

Anonymous

Mr. Meant-To has a comrade,
And his name is Didn't-Do.
Have you ever chanced to meet them?
Did they ever call on you?

These two fellows live together,
In the house of Never-Win.
And I'm told that it is haunted,
By the ghost of Might-Have-Been.

These two fellows

live together,

In the house of

Never-Win.

Mr. Meant-To

Anonymous

Mr. Meant-To has a comrade,
And his name is Didn't-Do.
Have you ever chanced to meet them?
Did they ever call on you?

These two fellows live together,
In the house of Never-Win.
And I'm told that it is haunted,
By the ghost of Might-Have-Been.

And I'm told that

it is haunted,

By the ghost of

Might-Have-Been.

January Brings the Snow

Sara Coleridge, 1834

January brings the snow,
Makes our feet and fingers glow.

February brings the rain,
Thaws the frozen lake again.

March brings breezes sharp and chill,
Shakes the dancing daffodil.

April brings the primrose sweet,
Scatters daisies at our feet.

May brings flocks of pretty lambs,
Sporting round their fleecy dams.

June brings tulips, lilies, roses,
Fills the children's hands with posies.

Hot July brings thunder showers,
Apricots and gillyflowers.

August brings the ears of corn,
Then the harvest home is borne.

Warm September brings the fruit,
Sportsmen then begin to shoot.

Brown October brings the pheasant,
Then to gather nuts is pleasant.

Dull November brings the blast,
Hark ! the leaves are whirling fast.

Cold December brings the sleet,
Blazing fire and Christmas treat.

January brings
the snow,
Makes our feet
and fingers glow.

January Brings the Snow

Sara Coleridge, 1834

January brings the snow,
Makes our feet and fingers glow.

February brings the rain,
Thaws the frozen lake again.

March brings breezes sharp and chill,
Shakes the dancing daffodil.

April brings the primrose sweet,
Scatters daisies at our feet.

May brings flocks of pretty lambs,
Sporting round their fleecy dams.

June brings tulips, lilies, roses,
Fills the children's hands with posies.

Hot July brings thunder showers,
Apricots and gillyflowers.

August brings the ears of corn,
Then the harvest home is borne.

Warm September brings the fruit,
Sportsmen then begin to shoot.

Brown October brings the pheasant,
Then to gather nuts is pleasant.

Dull November brings the blast,
Hark ! the leaves are whirling fast.

Cold December brings the sleet,
Blazing fire and Christmas treat.

February brings the rain, Thaws the frozen lake again.

January Brings the Snow

Sara Coleridge, 1834

January brings the snow,
Makes our feet and fingers glow.

February brings the rain,
Thaws the frozen lake again.

March brings breezes sharp and chill,
Shakes the dancing daffodil.

April brings the primrose sweet,
Scatters daisies at our feet.

May brings flocks of pretty lambs,
Sporting round their fleecy dams.

June brings tulips, lilies, roses,
Fills the children's hands with posies.

Hot July brings thunder showers,
Apricots and gillyflowers.

August brings the ears of corn,
Then the harvest home is borne.

Warm September brings the fruit,
Sportsmen then begin to shoot.

Brown October brings the pheasant,
Then to gather nuts is pleasant.

Dull November brings the blast,
Hark! the leaves are whirling fast.

Cold December brings the sleet,
Blazing fire and Christmas treat.

March brings breezes sharp and chill, Shakes the dancing daffodil.

January Brings the Snow

Sara Coleridge, 1834

January brings the snow,
Makes our feet and fingers glow.

February brings the rain,
Thaws the frozen lake again.

March brings breezes sharp and chill,
Shakes the dancing daffodil.

April brings the primrose sweet,
Scatters daisies at our feet.

May brings flocks of pretty lambs,
Sporting round their fleecy dams.

June brings tulips, lilies, roses,
Fills the children's hands with posies.

Hot July brings thunder showers,
Apricots and gillyflowers.

August brings the ears of corn,
Then the harvest home is borne.

Warm September brings the fruit,
Sportsmen then begin to shoot.

Brown October brings the pheasant,
Then to gather nuts is pleasant.

Dull November brings the blast,
Hark ! the leaves are whirling fast.

Cold December brings the sleet,
Blazing fire and Christmas treat.

April brings the
primrose sweet,
Scatters daisies
at our feet.

January Brings the Snow

Sara Coleridge, 1834

January brings the snow,
Makes our feet and fingers glow.

February brings the rain,
Thaws the frozen lake again.

March brings breezes sharp and chill,
Shakes the dancing daffodil.

April brings the primrose sweet,
Scatters daisies at our feet.

May brings flocks of pretty lambs,
Sporting round their fleecy dams.

June brings tulips, lilies, roses,
Fills the children's hands with posies.

Hot July brings thunder showers,
Apricots and gillyflowers.

August brings the ears of corn,
Then the harvest home is borne.

Warm September brings the fruit,
Sportsmen then begin to shoot.

Brown October brings the pheasant,
Then to gather nuts is pleasant.

Dull November brings the blast,
Hark! the leaves are whirling fast.

Cold December brings the sleet,
Blazing fire and Christmas treat.

May brings flocks

of pretty lambs,

Sporting round

their fleecy dams.

January Brings the Snow

Sara Coleridge, 1834

January brings the snow,
Makes our feet and fingers glow.

February brings the rain,
Thaws the frozen lake again.

March brings breezes sharp and chill,
Shakes the dancing daffodil.

April brings the primrose sweet,
Scatters daisies at our feet.

May brings flocks of pretty lambs,
Sporting round their fleecy dams.

June brings tulips, lilies, roses,
Fills the children's hands with posies.

Hot July brings thunder showers,
Apricots and gillyflowers.

August brings the ears of corn,
Then the harvest home is borne.

Warm September brings the fruit,
Sportsmen then begin to shoot.

Brown October brings the pheasant,
Then to gather nuts is pleasant.

Dull November brings the blast,
Hark ! the leaves are whirling fast.

Cold December brings the sleet,
Blazing fire and Christmas treat.

June brings tulips,

lilies, roses, Fills

the children's hands

with posies.

January Brings the Snow

Sara Coleridge, 1834

January brings the snow,
Makes our feet and fingers glow.

February brings the rain,
Thaws the frozen lake again.

March brings breezes sharp and chill,
Shakes the dancing daffodil.

April brings the primrose sweet,
Scatters daisies at our feet.

May brings flocks of pretty lambs,
Sporting round their fleecy dams.

June brings tulips, lilies, roses,
Fills the children's hands with posies.

Hot July brings thunder showers,
Apricots and gillyflowers.

August brings the ears of corn,
Then the harvest home is borne.

Warm September brings the fruit,
Sportsmen then begin to shoot.

Brown October brings the pheasant,
Then to gather nuts is pleasant.

Dull November brings the blast,
Hark! the leaves are whirling fast.

Cold December brings the sleet,
Blazing fire and Christmas treat.

Hot July brings

thunder showers,

Apricots and

gillyflowers.

January Brings the Snow

Sara Coleridge, 1834

January brings the snow,
Makes our feet and fingers glow.

February brings the rain,
Thaws the frozen lake again.

March brings breezes sharp and chill,
Shakes the dancing daffodil.

April brings the primrose sweet,
Scatters daisies at our feet.

May brings flocks of pretty lambs,
Sporting round their fleecy dams.

June brings tulips, lilies, roses,
Fills the children's hands with posies.

Hot July brings thunder showers,
Apricots and gillyflowers.

August brings the ears of corn,
Then the harvest home is borne.

Warm September brings the fruit,
Sportsmen then begin to shoot.

Brown October brings the pheasant,
Then to gather nuts is pleasant.

Dull November brings the blast,
Hark ! the leaves are whirling fast.

Cold December brings the sleet,
Blazing fire and Christmas treat.

August brings the ears of corn, Then the harvest home is borne.

January Brings the Snow

Sara Coleridge, 1834

January brings the snow,
Makes our feet and fingers glow.

February brings the rain,
Thaws the frozen lake again.

March brings breezes sharp and chill,
Shakes the dancing daffodil.

April brings the primrose sweet,
Scatters daisies at our feet.

May brings flocks of pretty lambs,
Sporting round their fleecy dams.

June brings tulips, lilies, roses,
Fills the children's hands with posies.

Hot July brings thunder showers,
Apricots and gillyflowers.

August brings the ears of corn,
Then the harvest home is borne.

Warm September brings the fruit,
Sportsmen then begin to shoot.

Brown October brings the pheasant,
Then to gather nuts is pleasant.

Dull November brings the blast,
Hark ! the leaves are whirling fast.

Cold December brings the sleet,
Blazing fire and Christmas treat.

Warm September

brings the fruit,

Sportsmen then

begin to shoot.

January Brings the Snow

Sara Coleridge, 1834

January brings the snow,
Makes our feet and fingers glow.

February brings the rain,
Thaws the frozen lake again.

March brings breezes sharp and chill,
Shakes the dancing daffodil.

April brings the primrose sweet,
Scatters daisies at our feet.

May brings flocks of pretty lambs,
Sporting round their fleecy dams.

June brings tulips, lilies, roses,
Fills the children's hands with posies.

Hot July brings thunder showers,
Apricots and gillyflowers.

August brings the ears of corn,
Then the harvest home is borne.

Warm September brings the fruit,
Sportsmen then begin to shoot.

Brown October brings the pheasant,
Then to gather nuts is pleasant.

Dull November brings the blast,
Hark ! the leaves are whirling fast.

Cold December brings the sleet,
Blazing fire and Christmas treat.

Brown October

brings the pheasant,

Then to gather

nuts is pleasant.

January Brings the Snow

Sara Coleridge, 1834

January brings the snow,
Makes our feet and fingers glow.

February brings the rain,
Thaws the frozen lake again.

March brings breezes sharp and chill,
Shakes the dancing daffodil.

April brings the primrose sweet,
Scatters daisies at our feet.

May brings flocks of pretty lambs,
Sporting round their fleecy dams.

June brings tulips, lilies, roses,
Fills the children's hands with posies.

Hot July brings thunder showers,
Apricots and gillyflowers.

August brings the ears of corn,
Then the harvest home is borne.

Warm September brings the fruit,
Sportsmen then begin to shoot.

Brown October brings the pheasant,
Then to gather nuts is pleasant.

Dull November brings the blast,
Hark ! the leaves are whirling fast.

Cold December brings the sleet,
Blazing fire and Christmas treat.

Dull November

brings the blast,

Hark! the leaves

are whirling fast.

January Brings the Snow

Sara Coleridge, 1834

January brings the snow,
Makes our feet and fingers glow.

February brings the rain,
Thaws the frozen lake again.

March brings breezes sharp and chill,
Shakes the dancing daffodil.

April brings the primrose sweet,
Scatters daisies at our feet.

May brings flocks of pretty lambs,
Sporting round their fleecy dams.

June brings tulips, lilies, roses,
Fills the children's hands with posies.

Hot July brings thunder showers,
Apricots and gillyflowers.

August brings the ears of corn,
Then the harvest home is borne.

Warm September brings the fruit,
Sportsmen then begin to shoot.

Brown October brings the pheasant,
Then to gather nuts is pleasant.

Dull November brings the blast,
Hark ! the leaves are whirling fast.

Cold December brings the sleet,
Blazing fire and Christmas treat.

Cold December

brings the sleet,

Blazing fire and

Christmas treat.

Frogs At School

George Cooper

Twenty froggies went to school,
Down beside a rushy pool;
Twenty little coats of green,
Twenty vests all white and clean.

"We must be on time," said they,
"First we study, then we play.
That is how we keep the rule,
When we froggies go to school."

Master Bullfrog, grave and stern,
Called the classes in their turn;
Taught them how to nobly strive;
Likewise how to leap and dive.

From his seat upon a log,
Showed them how to say "Kerchog"
Also how to dodge a blow
From the sticks which bad boys throw.

Twenty froggies grew up fast;
Bullfrogs they became at last;
Not one dunce was in the lot,
Not one lesson they forgot.

Polished in a high degree,
As each froggy ought to be,
Now they sit on other logs,
Teaching other little frogs.

Twenty froggies
went to school,
Down beside a
rushy pool;

Frogs At School

George Cooper

Twenty froggies went to school.
Down beside a rushy pool;
Twenty little coats of green,
Twenty vests all white and clean.

"We must be on time," said they,
"First we study, then we play.
That is how we keep the rule,
When we froggies go to school."

Master Bullfrog, grave and stern,
Called the classes in their turn;
Taught them how to nobly strive;
Likewise how to leap and dive.

From his seat upon a log,
Showed them how to say "Kerchog!"
Also how to dodge a blow
From the sticks which bad boys throw.

Twenty froggies grew up fast;
Bullfrogs they became at last;
Not one dunce was in the lot,
Not one lesson they forgot.

Polished in a high degree,
As each froggy ought to be,
Now they sit on other logs,
Teaching other little frogs.

Twenty little coats

of green,

Twenty vests all

white and clean.

Frogs At School

George Cooper

Twenty froggies went to school
Down beside a rushy pool;
Twenty little coats of green,
Twenty vests all white and clean.

"We must be on time," said they,
"First we study, then we play.
That is how we keep the rule,
When we froggies go to school."

Master Bullfrog, grave and stern,
Called the classes in their turn;
Taught them how to nobly strive;
Likewise how to leap and dive.

From his seat upon a log,
Showed them how to say "Kerchog!"
Also how to dodge a blow
From the sticks which bad boys throw.

Twenty froggies grew up fast;
Bullfrogs they became at last;
Not one dunce was in the lot,
Not one lesson they forgot.

Polished in a high degree,
As each froggy ought to be,
Now they sit on other logs,
Teaching other little frogs.

"We must be on
time," said they,
"First we study,
then we play.

Frogs At School

George Cooper

Twenty froggies went to school,
Down beside a rushy pool;
Twenty little coats of green,
Twenty vests all white and clean.

"We must be on time," said they,
First we study, then we play.
That is how we keep the rule,
When we froggies go to school."

Master Bullfrog, grave and stern,
Called the classes in their turn;
Taught them how to nobly strive;
Likewise how to leap and dive.

From his seat upon a log,
Showed them how to say "Kerchog!"
Also how to dodge a blow
From the sticks which bad boys throw.

Twenty froggies grew up fast;
Bullfrogs they became at last;
Not one dunce was in the lot,
Not one lesson they forgot.

Polished in a high degree,
As each froggy ought to be,
Now they sit on other logs,
Teaching other little frogs.

That is how we

keep the rule,

When we froggies

go to school."

Frogs At School

George Cooper

Twenty froggies went to school,
Down beside a rushy pool;
Twenty little coats of green,
Twenty vests all white and clean.

"We must be on time," said they,
"First we study, then we play.
That is how we keep the rule,
When we froggies go to school."

Master Bullfrog, grave and stern,
Called the classes in their turn;
Taught them how to nobly strive,
Likewise how to leap and dive.

From his seat upon a log,
Showed them how to say "Kerchog!"
Also how to dodge a blow
From the sticks which bad boys throw.

Twenty froggies grew up fast;
Bullfrogs they became at last;
Not one dunce was in the lot,
Not one lesson they forgot.

Polished in a high degree,
As each froggy ought to be,
Now they sit on other logs,
Teaching other little frogs.

Master Bullfrog,
grave and stern,
Called the classes
in their turn;

Frogs At School

George Cooper

Twenty froggies went to school,
Down beside a rushy pool;
Twenty little coats of green,
Twenty vests all white and clean.

"We must be on time," said they,
"First we study, then we play.
That is how we keep the rule,
When we froggies go to school."

Master Bullfrog, grave and stern,
Called the classes in their turn;
Taught them how to nobly strive;
Likewise how to leap and dive.

From his seat upon a log,
Showed them how to say "Kerchog!"
Also how to dodge a blow
From the sticks which bad boys throw.

Twenty froggies grew up fast;
Bullfrogs they became at last;
Not one dunce was in the lot,
Not one lesson they forgot.

Polished in a high degree,
As each froggy ought to be,
Now they sit on other logs,
Teaching other little frogs.

Taught them how

to nobly strive;

Likewise how to

leap and dive.

Frogs At School

George Cooper

Twenty froggies went to school,
Down beside a rushy pool;
Twenty little coats of green,
Twenty vests all white and clean.

"We must be on time," said they,
"First we study, then we play.
That is how we keep the rule,
When we froggies go to school."

Master Bullfrog, grave and stern,
Called the classes in their turn;
Taught them how to nobly strive,
Likewise how to leap and dive.

From his seat upon a log,
Showed them how to say "Kerchog!"
Also how to dodge a blow
From the sticks which bad boys throw.

Twenty froggies grew up fast;
Bullfrogs they became at last;
Not one dunce was in the lot,
Not one lesson they forgot.

Polished in a high degree,
As each froggy ought to be,
Now they sit on other logs,
Teaching other little frogs.

From his seat

upon a log,

Showed them how

to say "Kerchog!"

Frogs At School

George Cooper

Twenty froggies went to school,
Down beside a rushy pool;
Twenty little coats of green,
Twenty vests all white and clean.

"We must be on time," said they,
"First we study, then we play.
That is how we keep the rule,
When we froggies go to school."

Master Bullfrog, grave and stern,
Called the classes in their turn;
Taught them how to nobly strive;
Likewise how to leap and dive.

From his seat upon a log,
Showed them how to say "Kerchog!"
Also how to dodge a blow
From the sticks which bad boys throw.

Twenty froggies grew up fast;
Bullfrogs they became at last;
Not one dunce was in the lot,
Not one lesson they forgot.

Polished in a high degree,
As each froggy ought to be,
Now they sit on other logs,
Teaching other little frogs.

Also how to dodge

a blow

From the sticks

which bad boys

throw.

Frogs At School

George Cooper

Twenty froggies went to school,
Down beside a rushy pool;
Twenty little coats of green,
Twenty vests all white and clean.

"We must be on time," said they,
"First we study, then we play.
That is how we keep the rule,
When we froggies go to school."

Master Bullfrog, grave and stern,
Called the classes in their turn;
Taught them how to nobly strive;
Likewise how to leap and dive.

From his seat upon a log,
Showed them how to say "Kerchog!"
Also how to dodge a blow
From the sticks which bad boys throw.

Twenty froggies grew up fast;
Bullfrogs they became at last;
Not one dunce was in the lot,
Not one lesson they forgot.

Polished in a high degree,
As each froggy ought to be,
Now they sit on other logs,
Teaching other little frogs.

Tweny froggies

grew up fast;

Bullfrogs they

became at last;

Frogs At School

George Cooper

Twenty froggies went to school,
Down beside a rushy pool;
Twenty little coats of green,
Twenty vests all white and clean.

"We must be on time," said they,
"First we study, then we play,
That is how we keep the rule,
When we froggies go to school."

Master Bullfrog, grave and stern,
Called the classes in their turn;
Taught them how to nobly strive;
Likewise how to leap and dive.

From his seat upon a log,
Showed them how to say "Kerchog!"
Also how to dodge a blow
From the sticks which bad boys throw.

Tweny froggies grew up fast;
Bullfrogs they became at last;
Not one dunce was in the lot,
Not one lesson they forgot.

Polished in a high degree,
As each froggy ought to be,
Now they sit on other logs,
Teaching other little frogs.

Not one dunce was
in the lot,
Not one lesson
they forgot.

Frogs At School

George Cooper

Twenty froggies went to school,
Down beside a rushy pool;
Twenty little coats of green,
Twenty vests all white and clean.

"We must be on time," said they,
"First we study, then we play,
That is how we keep the rule,
When we froggies go to school."

Master Bullfrog, grave and stern,
Called the classes in their turn;
Taught them how to nobly strive;
Likewise how to leap and dive.

From his seat upon a log,
Showed them how to say "Kerchog!"
Also how to dodge a blow
From the sticks which bad boys throw.

Tweny froggies grew up fast;
Bullfrogs they became at last;
Not one dunce was in the lot,
Not one lesson they forgot.

Polished in a high degree,
As each froggy ought to be,
Now they sit on other logs,
Teaching other little frogs.

Polished in a high

degree,

As each froggy

ought to be,

Frogs At School

George Cooper

Twenty froggies went to school,
Down beside a rushy pool;
Twenty little coats of green,
Twenty vests all white and clean.

"We must be on time," said they,
"First we study, then we play,
That is how we keep the rule,
When we froggies go to school."

Master Bullfrog, grave and stern,
Called the classes in their turn;
Taught them how to nobly strive;
Likewise how to leap and dive.

From his seat upon a log,
Showed them how to say "Kerchog!"
Also how to dodge a blow
From the sticks which bad boys throw.

Tweny froggies grew up fast;
Bullfrogs they became at last;
Not one dunce was in the lot,
Not one lesson they forgot.

Polished in a high degree,
As each froggy ought to be,
Now they sit on other logs,
Teaching other little frogs.

Now they sit on

other logs,

Teaching other

little frogs.